Fossils Tell of Long Ago

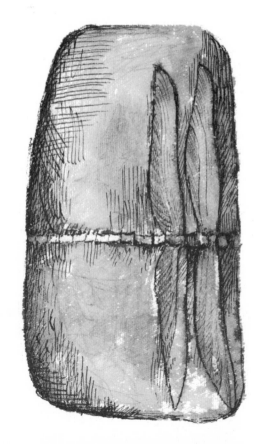

A Let's-Read-and-Find-Out Book ™

Fossils Tell of Long Ago

BY ALIKI

A Harper Trophy Book
HARPER & ROW, PUBLISHERS

LET'S-READ-AND-FIND-OUT BOOKS™

The *Let's-Read-and-Find-Out Book*™ series was originated by Dr. Franklyn M. Branley, Astronomer Emeritus and former Chairman of the American Museum-Hayden Planetarium, and was formerly co-edited by him and Dr. Roma Gans, Professor Emeritus of Childhood Education, Teachers College, Columbia University. Text and illustrations for each of the more than 100 books in the series are checked for accuracy by an expert in the relevant field. Titles available in paperback are listed below. Look for them at your local bookstore or library.

Air Is All Around You · Ant Cities · A Baby Starts to Grow · The BASIC Book
Bees and Beelines · The Beginning of the Earth · Bits and Bytes · Comets · Corn Is Maize
Danger—Icebergs! · Digging Up Dinosaurs · Dinosaurs Are Different · A Drop of Blood
Ducks Don't Get Wet · Eclipse · Fireflies in the Night · Flash, Crash, Rumble, and Roll
Fossils Tell of Long Ago · Germs Make Me Sick! · Get Ready for Robots! · Glaciers
Gravity Is a Mystery · Hear Your Heart · How a Seed Grows · How Many Teeth?
How to Talk to Your Computer · Hurricane Watch · Is There Life in Outer Space?
Journey into a Black Hole · Look at Your Eyes · Me and My Family Tree · Meet the Computer
The Moon Seems to Change · My Five Senses · My Visit to the Dinosaurs
No Measles, No Mumps for Me · Oxygen Keeps You Alive · The Planets in Our Solar System
Rock Collecting · Rockets and Satellites · The Skeleton Inside You · The Sky Is Full of Stars
Snow Is Falling · Straight Hair, Curly Hair · The Sun: Our Nearest Star · Sunshine Makes the Seasons
A Tree Is a Plant · Turtle Talk · Volcanoes · Water for Dinosaurs and You · What Happens to a Hamburger
What I Like About Toads · What Makes Day and Night · What The Moon Is Like · Why Frogs Are Wet
Wild and Woolly Mammoths · Your Skin and Mine

Fossils Tell of Long Ago
Copyright © 1972 by Aliki Brandenberg
Library of Congress Catalog Card Number: 85-42977
Trade ISBN 0-690-31378-0
Library ISBN 0-690-31379-9
Trophy ISBN 0-06-445004-X

Published in hardcover by Thomas Y. Crowell, New York.

33045

Fossils Tell of Long Ago

Once upon a time a huge fish was swimming around
   when along came a smaller fish.
The big fish was so hungry it swallowed the other
   fish whole.
The big fish died and sank to the bottom of the lake.

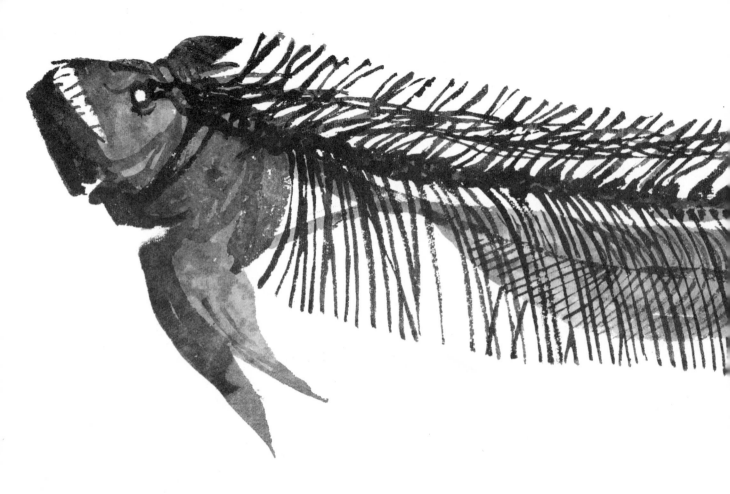

This happened ninety million years ago.

How do we know?

We know because the fish turned to stone.

The fish became a fossil.

A plant or an animal that has turned to stone is called a fossil.

Scientists can tell how old stones are.
They could tell how old the fish fossil is.
So we know how long ago the fish lived.

How did plants and animals become fossils?
Most plants and animals do not become fossils when they die.
They rot,

or they crumble, dry up, and blow away.
No trace of them is left.

This could have happened to the big fish.
We would never know it had lived.
Instead, the fish became a fossil.
This is how it happened.

When the big fish died, it sank into the mud at the
  bottom of the lake.
Slowly, the fish rotted.
Only its bones were left.
The bones of the fish it had eaten were left, too.
The skeleton of the fish lay buried and protected deep
  in the mud.

Thousands of years went by.
More and more mud covered the fish.
Tons and tons of mud piled up.
After a long time, the surface of the earth changed.
The lake in which the fish was buried dried out.

It rained on the drying mud.

Water seeped through the mud.

Minerals from stones were dissolved in the water.

The water seeped into all the tiny holes in the fish bones.

The minerals in the water were left behind in the fish bones.

After a very long time the minerals changed the bones
    to stone.

The fish was a fossil.

The mud around the bones became hard as rock, too.

Some fossils, like the fish, are bones or shells that
  have turned to stone.
Sometimes a fossil is only an imprint of a plant or an animal.

Millions of years ago a fern grew in a forest.
It fell and was buried in swampy ground.

The fern rotted away.
But it left the mark of its shape in the mud.
It left its imprint.
The mud hardened.
The mud, with the imprint of the fern,
  became a fossil called coal.
Many fossils of plants and animals are found in coal.

This is a dinosaur track.
It was made in fresh mud two hundred million years ago.

Hot melted stone from a volcano filled the dinosaur's
 footprint in the mud.
The stone cooled and hardened.
A few years ago fossil hunters dug through the stone.
They found an exact imprint of a dinosaur's foot.

Not all fossils are found in stone.
Some are found in the frozen ground of the Arctic.
This ancient mammoth was a kind of elephant.
It was frozen thousands of years ago.
It was found not long ago buried in the frozen ground.
The grass it had been eating was still in its mouth.

The mammoth was fresh enough to eat!
Someone who ate a piece
    said it was dry and not very tasty.
But what could you expect from an ancient mammoth?

15

Millions of years ago a fly was caught in the sticky
  sap of a pine tree.
The sap hardened and became a fossil called amber.
Amber looks like yellow glass.
The fly was perfectly preserved in the amber.

FLY

Other insects and even plants have been preserved in amber.

FERN LEAF

SPIDER

COCKROACH

We have learned many things from the fish, the fern,
   the fly, and the dinosaur track.
Fossils tell us about the past.
Fossils tell us there once were forests—

where now there are rivers.
We find fossils of trees in the bottoms of some riverbeds.

Fossils tell us there once were seas where now there
   are mountains.
Fossils of sea plants and animals have been found on
   mountains.

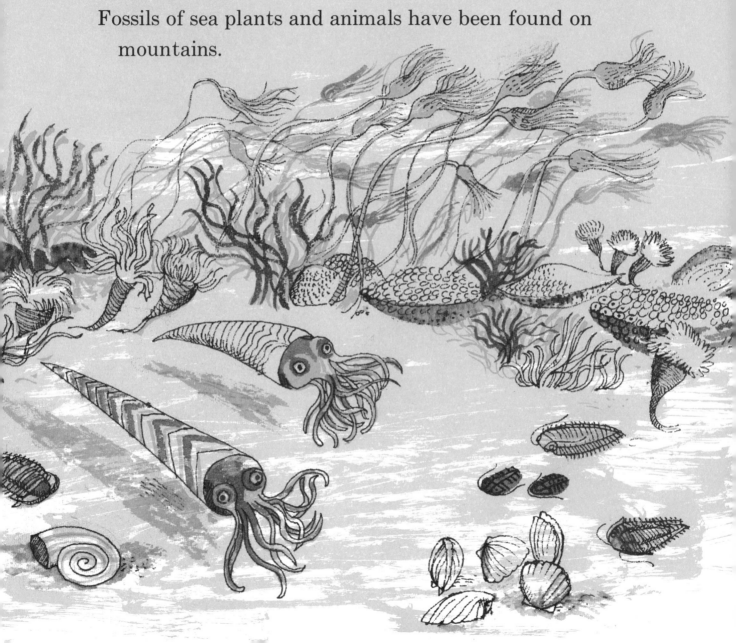

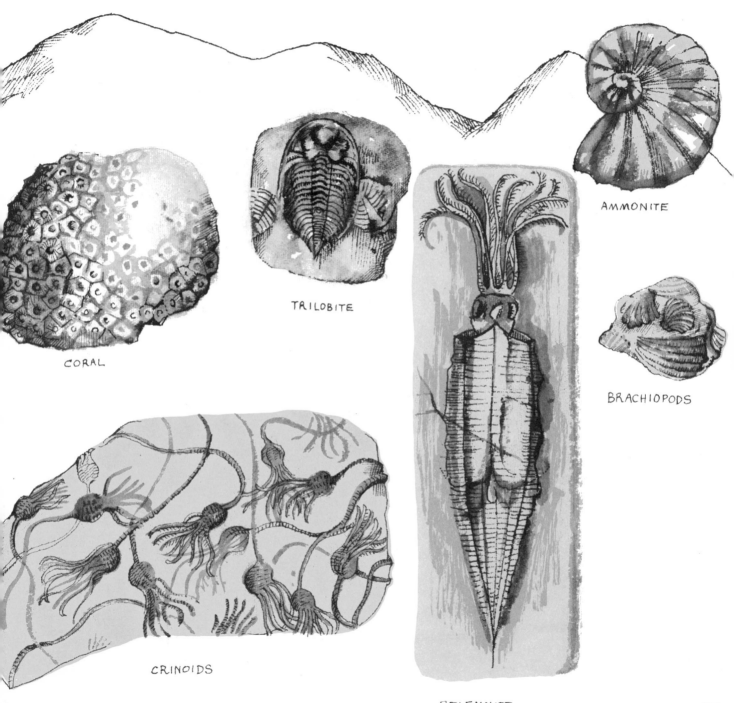

CORAL

TRILOBITE

AMMONITE

BRACHIOPODS

CRINOIDS

BELEMNITE

21

Many lands that are cold today were once warm.
We find fossils of tropical plants in very cold places.

Fossils tell us about strange creatures that lived on
   earth long ago.
They tell us about dinosaurs,

TYRANNOSAURUS

STEGOSAURUS

pteranodons and ichthyosaurs.
No such creatures are alive today.
They have died out.
We say they are extinct.

Some fossils are found by accident.
Some are found by fossil hunters who dig for them.

FOSSIL HUNTERS DIGGING OUT THE FOSSIL
OF THE BIG FISH, PORTHEUS, IN KANSAS.

You, too, might find a fossil if you look hard.
When you see a stone, look at it carefully.
It may be a fossil of something that once lived.
You may find a fossil at the seashore.

You may find a fossil in the woods, or by a newly dug road.
You may find a fossil in the field or on a mountain top.
If you live in the city, you may find a fossil there, too.
Sometimes you can see them in the polished limestone
walls of some buildings.

How would you like to make a fossil?
Not a one-million-year-old fossil but a
    one-minute-old "fossil."
Make a clay imprint of your hand, like this:

Take some clay.
Flatten it out.
Press your hand in the clay.
Lift your hand away.

Your hand is gone, but its shape is in the clay.
You made an imprint.
The imprint shows what your hand is like, the way
    a dinosaur's track shows us what his foot was like.

Suppose, when it dried, you buried your clay imprint.
Suppose, a million years from now, someone found it.
Your imprint would be hard as stone.
It would be a fossil of your hand.
It would tell something about you.
It would tell the finder something about life on earth
   a million years earlier.

Every time anyone finds a fossil we learn more about
   life on earth long ago.
Someday you may find a fossil, one that is millions
   and millions of years old.
You may discover something that no one knows today.

ABOUT ALIKI

One of the things Aliki likes to do best, along with drawing pictures and writing books and growing plants, is to travel.

One summer Aliki and her husband, Franz Brandenberg, and their children, Jason and Alexa, were in Greece. It was there, on a dusty road, that Jason found a fossil of brachiopods.

Although Aliki had been interested in fossils before then, this book is a result of Jason's discovery.

Aliki Brandenberg grew up in Philadelphia and was graduated from the Philadelphia College of Art. She worked in many phases of art before she began writing and illustrating books for children.